Animal Groups

Fish

By Dalton Rains

www.littlebluehousebooks.com

Little Blue House is distributed by North Star Editions:
sales@northstareditions.com | 888-417-0195

Produced for Little Blue House by Red Line Editorial.

Photographs ©: Shutterstock Images, cover, 4, 7, 9, 11, 13, 15, 17, 19, 21, 22–23 (water background), 22–23 (anatomy of fish), 24 (top left), 24 (top right), 24 (bottom left), 24 (bottom right)

Library of Congress Control Number: 2022919920

ISBN
978-1-64619-808-5 (hardcover)
978-1-64619-837-5 (paperback)
978-1-64619-894-8 (ebook pdf)
978-1-64619-866-5 (hosted ebook)

Printed in the United States of America
Mankato, MN
082023

About the Author

Dalton Rains writes and edits nonfiction children's books. He lives in Minnesota.

Table of Contents

Fish 5

Is It a Fish? 22

Glossary 24

Index 24

tail

Fish

I see a fish.

It has a tail.

I see a fish.

It has gills.

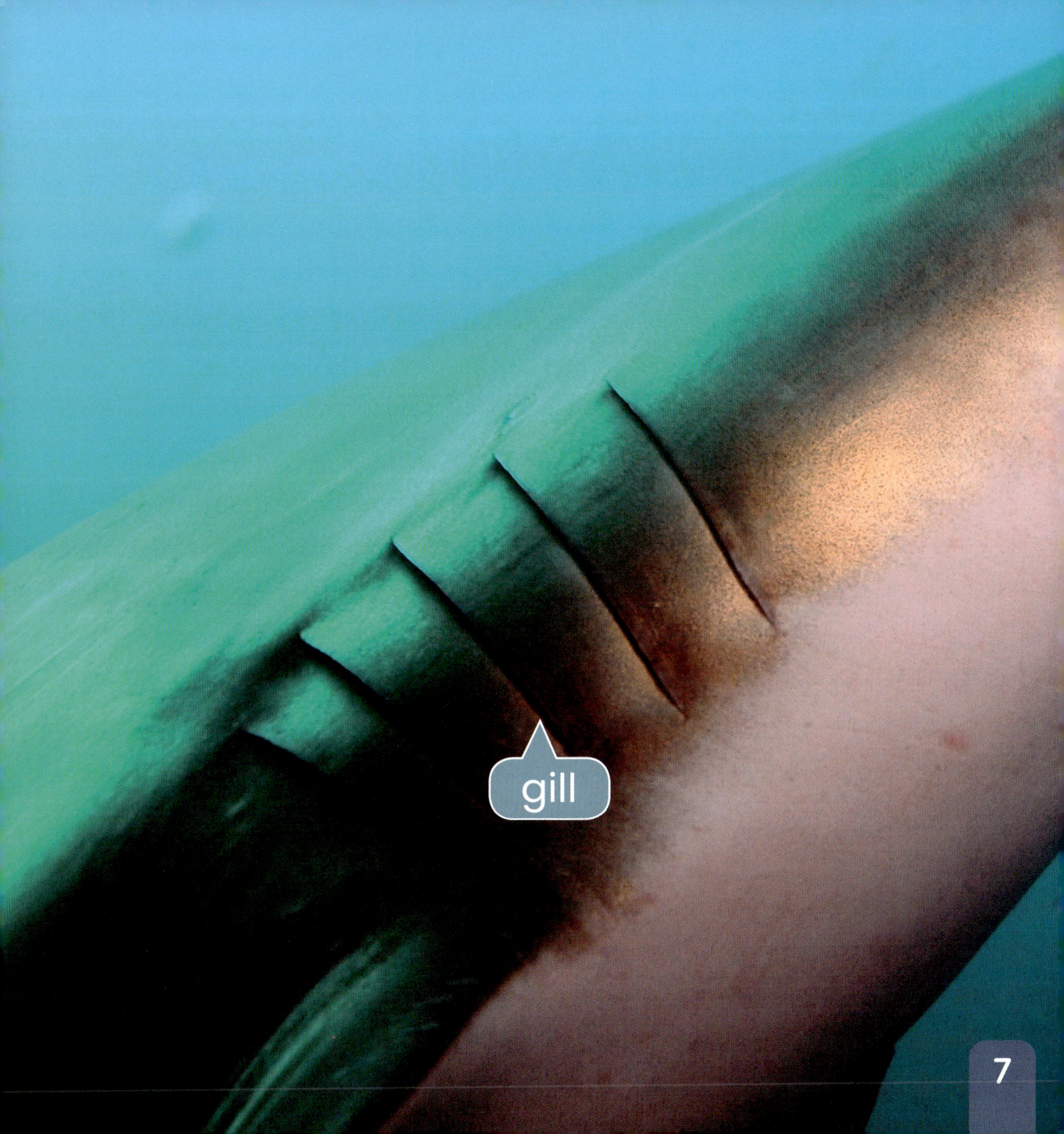
gill

I see a fish.

It has scales.

scales

I see a fish.

It has a mouth.

mouth

I see a fish.

It has teeth.

tooth

I see a fish.

It has eyes.

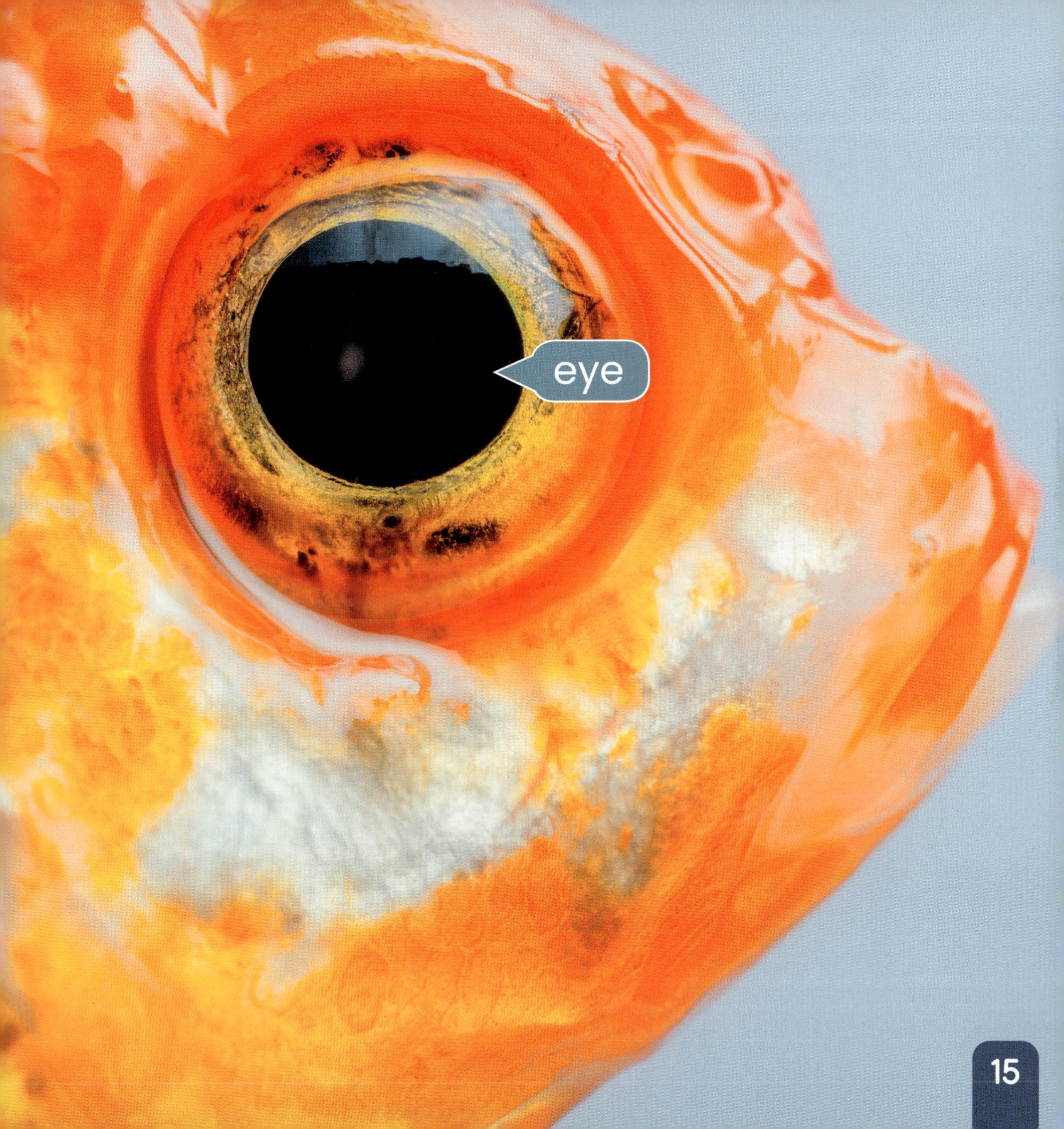
eye

I see a fish.

It has a fin.

fin

I see a fish.

It has babies.

baby

I see a fish.

It has stripes.

stripe

Is It a Fish?

All fish have gills.

All fish live in the water.

All fish have backbones.
All fish have fins.

Glossary

fin

scales

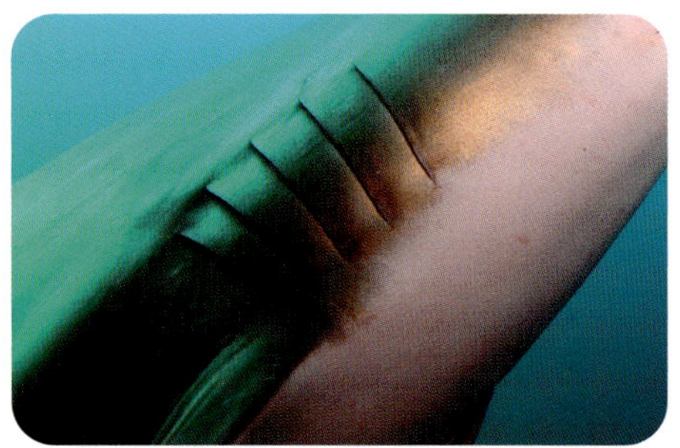
gills

stripes

Index

B
babies, 18

E
eyes, 14

M
mouth, 10

T
teeth, 12